Synchronicity and the Secret of the One

Published by Robert Torres – Real Time Records Publishing
All rights reserved. No part of this book may be reproduced or transmitted in any form or by any means, electronic or mechanical, including photocopying, recording or by any information storage and retrieval system, without written permission from the author, except for the inclusion of brief quotations in a review.
Copyright © 2018 Robert Torres:
Published in the United States of America

Special Thanks

I'd like to thank everyone traveling with me during this strange journey. Especially my family and friends with an extra thanks to the wonderful Melanie Torres, Robert, Janet Puccino, Michael Negron, Taylor, Jina Puccino and the rest of the bunch. Robert, Saige and Aiden. All the great scientists, theologians, philosophers, authors, beings, entities, the supreme being of course, the universe, the planet, other gods and anyone I left out. Cover art by Robert Torres.

Table of Contents

About the Author – 82

Contact -81

Preface -A

Introduction -AA

Ch. 1 --- Synchronicity and Other Strange Things ---1
Ch. 2 --- My First Experiences --- 9
Ch. 3 --- Language, Creation and Evolution --- 14
Ch. 4 --- The Dilemma: Knowledge – Belief --- 25
Ch. 5 --- The Science Dilemma --- 35
Ch. 6 --- The Philosophy Dilemma --- 43
Ch. 7 --- Dimensions of Consciousness / Meaning --- 49
Ch. 8 --- "The Flow" --- 55
Ch. 9 --- Determinism vs Chance --- 62
Ch. 10 --- Some Interesting Synchronicities --- 67
Ch. 11 --- Conclusion What About the Numbers - 75

Resources --- 79

Preface –A

Synchronicity and the Secret of the One is based on personal experiene and the idea that there is a force that is omnipresent throughout the universe that is so mysterious that as of yet science, philisophy, physics, nothing can explain it. And even though it can't be explained right now with our current knowledge. It is likely we are unable to see it because we are looking at the problem from too many logic based angles something for which there is no logic – no "one" accepted science. This is because when we are talking about certain subjects like synchronicity we are crossing over into the realm of the supernatural. These are some of the closely kept secrets of the universe. Magic, ghosts, UFOs, the oddities of the mind and all those things that fascinate us or make your hairs stand up. Do you believe your own eyes or rather - what you see? Most people do believe what they see and beyond that they don't believe what they don't see for themselves. Everything must be proven and on what do we base that proof? Ultimately we base it on an accepted human consensus which we base on some agreed upon litmus test. We agree when thousands of us experience the same thing or millions. Synchronicity is one of those things… and I experienced it many times.

Introduction -AA

The actual subject matter of Synchronicity although experienced by probably everyone is not accepted by almost anyone in the mainstream scientific community. But it's the exact type of reader I'm writing for. People with mysticism in their hearts mixed with some science curiosity in their minds. When you're done reading it I hope you'll grasp it as an eye opening and educational short fun science experience. Relizing some of what all so many of us us already know and have experienced it. I myself have experienced the paranormal many times and in big ways. Like most of you I could never understand it but I knew it was happening to many people and I wanted to know why. It was when I began looking for answers that my long journey inward began. I researched for years at science, magic, belief and many other systems for my article writing. I put together some examples of my own as well as some others as you will see in the coming pages. These are the actual events that I hope you enjoy reading.

Chapter 1
Synchronicity and Other Strange Things

Everywhere I look I find mystery and the further I look the more I learn. How do you research something for which everyone to this day after thousands of years has only theories for, or dismiss it outright? So I write about it because I lived it and I'll share some actual events that were personal to me. I am a believer in something that I don't know what it is. Crazy right? I'm trying to use actual science from my many years of researching this that I have put into my writing and let me tell you that writing the things I do has also been some experience.

Synchronicity and the Secret of the ONEs
11:11

The biggest secret is often times found right in front of you where it has always been – a contradiction in plain sight and not hidden at all – it's part of your world – a world of contradictions. It seems so strange that no matter where we look we find strangeness and even when we aren't paying attention it sometimes pushes right out at some of us. "Synchronicity is an ever present reality for those who have eyes to see" --- Carl Gustav Jung. But what does this mean?

It is one of the secrets hidden from even the greatest thinkers, scientists and theologians that have ever lived and so often because they can't see – they don't know how to see – they don't know where to look. They weren't taught to look... Most live in the reality of the three dimensional world we live in or rather the one we all know. The one where what "is" must have empirical proof to be real. Mysterious alignments just like 1111 or my number 3333 to most they're just numbers without context. But when we look at them with the deeper metaphysical understanding and knowledge they reveal the magic created by the universe itself. The magic within us that the universe reflects back at us like a mirror image. Only we have to be looking even when we aren't expecting anything we have to be open to accept what comes forth. Open your eyes at night to see the stars and open your eyes in the dark to see the wonder. Yes, you have to always have your eyes and mind open and ready to feel the wonder and to expect the unexpected…

So many of us see things so differently because so many of us don't see some things at all and only because we are too focused on sight when we should be using all of our senses. We live in a world full of sensations and we so often ignore most of them. Statistically meaningful synchronicities happen every second of every day to someone somewhere in the world. Yet we are told that these thing are not real or magical thinking. When they happen those of us who know that they are more than just coincidences may get a brief thrill. This is so often because of our attached perceptions coming forth from our individual subconscious connections. It is then by that act of "thought" alone that they become more than just random alignments because we make these connections. It's when we give them meaning that they receive meaning! This because we saw what the meaning was trying to show us perhaps. That's because for some personal reason we felt that they meant something more to us through these deeper inner connections. Then they become part of our real world by our own explicit projections. When we understand the realization that we are part of the creation process of reality. Our reality is added to the universal reality – the universal mind.

Synchronicity and the Secret of the One

To the skeptics or perhaps the level headed mathematician types this is often called confirmation bias (magical thinking). Or even wrong thinking because right thinking is to think – like everyone else who thinks normal – right? All good science must make sense but it must also be verifiably true? Only when we talk about synchronicity we are not talking about what makes sense or about what can be proven in a classroom or a lab. We are dimensions apart and I mean that literally. Multi-dimensional thinking is what we require when we enter these esoteric subjects and in fact it's the way this all works. This science was actually made up as it was going along. Not with proof, but with observations, deductions, speculation. Human observation is the science of knowing. "I know" because I saw it with my own eyes" is how mankind witnesses in order to pass our stories on to others. Tesla said: "Today's scientists have substituted mathematics for experiments, and they wander off through equation after equation, and eventually build a structure which has no relation to reality." Tesla was hands on – believing much of theoretical physics lacked, observed reality. It's the reason why we can read dozens of different theories on its nature with none of them being real.

For some of us the alignments of events activate some trigger. We are altered by triggers which in turn alter these subconscious or collective unconscious neural processes in our personal pre-material real world of the mind environments. What physicists David Bohm or David Peat call the sub-stratum that we can only hypothesize to be an unknown nontangible extra dimension. Meaning a yet unknown property of the known universe itself - creating these synchronicities. What either of them would call the "holomovement." A room full of people can be in the same place and see the same things and to one person it has significance. But it doesn't have to be one person because it could have significance to any number of people. As many times as I've had these experiences and what I can only call macro synchronicities it still fascinates me. For people like me it will always be this way. Fascination lives in my world because for some reason I am connected to this dimension of meaning. Everything fascinates me and much of it is because I base everything off of a foundation of metaphysics, science and wonder.

At the very beginning of my research I knew that for me everything started with several experiences with the paranormal and then just wanting to know why it happened to me or why does it happen to anyone. That led me to reading an amazing book by researcher and author of the paranormal, Charles Fort called the "Damned" published in 1915. The whole thing is about the information sidelined by mainstream media and brought to light by Fort through extensive investigative research. I was hooked – I ended up reading all his books! Charles Fort also tells many amazing stories in his books. Did you know he gained a cult following through his stories? Yes, they are called the Fortean Societies which have branches all over the world to this day and even a magazine called the Fortean Times. One section of his book "The Damned" he talks about poltergeists which he tells about the relationship between young girls. It seems paranormal investigators noticed a pattern having to do with many poltergeist hauntings and young troubled girls being seemingly present. Which was attributed to perhaps an inherent subconscious telekinesis. The first book of his I read "The Damned" which went through a lot of paranormal cases and kept it interesting.

Then it was on to Carl Gustav Jung and others with the same curiosity as those who began to build the science and language of parpsychology. I can recall one of the first stories I read on odd coincidences was from multiple sources. It was the story of a woman named Shelly which I'll retell here. "Shelley came from America and was sitting at Notre Dame in Paris giving her sore feet a rest. The shoes she had worn from the States had turned out to be painful, and her limited budget didn't allow her to buy another pair. Suddenly she felt an inner prompting, and she got up, walked out of the church, and turned left. Following her promptings, she made several other turns to arrive at a square. There, on top of a trash can, sat a pair of brand new black boots with no signs of wear — in exactly her size. "It was perfect," she said. "If they had been inside the trash can, I wouldn't have pulled them out. If they had been worn before, I wouldn't have put them on. And they were so stylish I never could have afforded them myself!" So is this an intuition story or a synchronicity story? Intuition got her to the boots. Synchronicity provided her with precisely what she needed – her size and all: she was virtually handed the boots by the Universe. Whatever it is, I liked the story. These are the stories that are hard to dismiss as coincidence.

It's like going back in time and lining it all up so that she would be there and before her for some reason someone had her size shoe and threw them out in the exact place she would walk. The time, the place and it boggles the mind when you dissect all that had to happen. Much of my experience when it comes to synchronicity also boggles the mind when I think of how things must align themselves with infinite regression. Newton's physics of "the clockwork universe" is one interpretation like his belief in God's universal clock set into motion from the beginning of time and left to run. Everything happens because it was all set into motion by the perfect creator or the universe and happens because of what preceded it or a "cause."

Like determinism or the belief that everything that happens was supposed to happen even individual human actions and therefore could happen no other way.

Chapter 2
My First Experiences

Synchronicity: *is the experience of two or more events that are apparently causally unrelated occurring together in a meaningful manner [to the observer]. To count as synchronicity, the events should be unlikely to occur together by chance.*

If you believe synchronicity is simply coincidence, then you haven't read any of the top experts in the field. The famous psychotherapist Carl Gustav Jung coined the term synchronicity in the 1920s to reference the alignment of universal forces with a person's experiences.

These forces have been sought out for centuries in many spiritual traditions as a means of aligning with the "flow." This usually takes years of disciplined meditation, study, ritual or by other means advancing through a graduated system to navigate this journey toward a harmonic "individuation." To some the search is inward for the self, yet for others it's an outward search for spirituality.

My first experience with synchronicity was on March 21st at 3:03am which is the 3rd month, 3rd week, 3rd hour, 3rd minute or 3333, on the equinox and the moment of my birth. It was my alignment with the universal forces, the planet, space and time.

My awakening also came about through no effort of my own by way of a supernatural encounter with an entity of light. It began my involvement with and research into the underlying spiritual and metaphysical nature of reality. Also in many different ways and on many different levels - I became "enlightened." Because of this I have an inherent understanding of the truth behind many esoteric concepts like the flow, the spirit, one-ness and even divinity.

My experiences of course are not entirely unique, but the way in which this window of understanding opened for me is. There was no journey, no explanation as to the reasons why – a light came down from the sky with no words at all and a certain amount of esoteric knowledge was just revealed to me. Things that to many are never more than concepts or metaphors I actually see as having real form. I've witnessed many paranormal phenomena unfold before me and with me seemingly being the only rational connection. However, because of my earlier encounters I've never attributed much of it directly to myself. I've always sensed a presence around me. So unlike many others who seek the path, I feel that for some reason the source found me.

Being human however always leaves me questioning: "Who or what is this presence?" Abilities like ESP, clairvoyance and telekinesis are all very real to me, along with the concepts of an underlying substratum or pre-spacetime otherwise known as pre-existence. Heavy stuff right – and but what is our connection?

In Jung's book "An Acausal Connecting Principle" it is subjective meaning that connects us. Without an observer (YOU) there is no mind, no synchronicity, no meaning. Thoughts connected to events, mind connected to movements of matter, absent of a cause (acausal). Thinking something before it happens, remote viewing, telekinesis, where do these abilities come from? Since the vast majority of mainstream scientists don't accept the mind as a cause. How then do we explain let alone prove any of this?

I say that we are using a flawed science because it is incomplete. Theoretical Physicist, Dr. William Tiller among others propose that consciousness is what's missing from the equation. It is the unifying integrator of all the individual constituents. Physicist David Bohm says there is a hidden variable implying that neither relativity nor quantum mechanics should be accepted as a conclusive nor exclusive solution. Jung's preference for subjective inner experience over empirical data opened him up to much criticism from many of his colleagues who asserted that he was countering materialism with mysticism. But after all this entire field falls within the very boundaries of parapsychology. Materialism is not required…

My research into synchronicity began with Jung, Pauli, Bohm and others but for millennia prior to Jung man has experienced synchronicity. Theoretically it begins outside of our space-time in the flow where all knowledge exists and our material reality takes shape. It then unfolds into our dimension of consciousness creating and gathering more information before returning back to the flow. Like the ripples on the surface of the stream which are never really separate from it. This is described by David Bohm's theory of "The Implicate and Explicate Order." However before there was the idea of synchronicity, humanity used different terms like correspondence, sympathy, harmony and unity. The "unus mundus" - the underlying unified reality from which everything emerges and to which everything returns.

Synchronicity and the Secret of the One

JUNG   PAULI   BOHM   HERACLITUS   HIPPOCRATES

It was in the fourth century B.C. that the Greek philosopher Heraclitus viewed all things as being interrelated - nothing was isolated and all things were linked. Similarly Hippocrates said: "There is ONE common flow, a common breathing. Everything is in sympathy." There is a bond - and even between inanimate objects. A form of "animism" or the belief that all matter has consciousness or a spiritual essence. This is a classic idea in early Greek philosophy whereby separateness is an illusion.

Chapter 3

Language, Creation and Evolution

Now what you should know is that Jung had a lifelong fascination in and many experiences with the paranormal. These experiences along with his research influenced all of his major theoretical formulations. They are also what led to his revolutionary ideas about synchronicity, which was an ever evolving hypothesis – lasting his entire life.

It all started with simple coincidences evolving further to include simultaneous compound or muli-coincidences, then coincidences separated by time. Then complexities began to emerge which some said he was fitting into his theory rather than accepting them as separate paranormal events. This led to what he referred to as "universal psychic structures" but which would later be called "archetypes." Archetypes were more simply, non-material universal patterns or structures (noumenon). These are common to all humanity and shaped by each of us individually. It is as he believed that our subconscious mind taps into the collective unconscious and helps create for each of us our own unique life experiences.

Building on these ideas he developed what he called the psychoid. Also simply put, this is a mixture of the all theoretical psychic dimension and the pre-existent materiality colluding behind the scenes to create our synchronicities. Developing the ideas of the "collective unconscious, individuation and the true self"

Eventually Jung's concepts would broaden to incorporate aspects of quantum physics. He cites his conversations with Einstein and his theory of relativity as his inspiration but it was his work with the Nobel Prize winning physicist Wolfgang Pauli that advanced his theories far beyond their simplistic beginnings. It was actually his aspects of the psychoid which so impressed Pauli who became convinced that parapsychology should be included into physics. The psycho-physical archetype (psychoid) was the link between mind and matter. This brought about much progress for Jung's hypotheses as well as Pauli's further understanding of the matrix. As they often borrowed concepts from each other. Jung would begin to incorporate the language and ideas of physics into his formations and vice versa with Pauli.

I find it amazing how it's mainstream science who make up the largest number of paranormal deniers as always and yet most of us who believe in the magic of the universe end up turning to mainstream science when searching for serious answers. Not because they have the answers, but because we give them credibility. Although Pauli however was himself a scientist he was also no stranger to the world of the strange. He as well was fascinated by the paranormal and had his own bouts with the supernatural. Enter his experiences with telekinesis. It has been written about that catastrophic breakdowns of expensive experimental equipment would inexplicably occur everywhere he went. Although it was often joked about, other scientists were said to have feared his presence during experiments. It was commonly believed and accepted that he was the cause experiments and equipment would blow up with no damage to himself. It was because of this that the experimental physicist Otto Stern banned Pauli from his laboratory in Hamburg, Germany despite being good friends, he didn't want Pauli anywhere near it.

Pauli's friend and colleague Rudolf Peierls described the Pauli Effect as follows: "This was a kind of spell he was supposed to cast on people or objects in his neighborhood, particularly in physics laboratories, causing accidents of all sorts. Machines would stop running when he arrived in a laboratory, a glass apparatus would suddenly break, a leak would appear in a vacuum system, but none of these accidents would ever hurt or inconvenience Pauli himself.

When important experimental equipment in Professor James Frank's laboratory at the Physics Institute at the University of Göttingen blew up for no apparent reason, someone remarked that this could be the Pauli effect. However, Pauli was nowhere in the area; he was on a train, traveling to Denmark. It was later discovered that at the time of the lab explosion, the train carrying Pauli from Zurich to Copenhagen was making a stop at Göttingen station.

When he arrived at Princeton in 1950, an expensive new cyclotron that had recently been installed burned for no obvious reason, and there was again speculation about the Pauli Effect.

Pauli was also convinced that what was happening around him was real and he got a kick out of it. This is now well known in physics as "The Pauli Effect." This was believed to be caused by psychic telekinesis (psychokinesis) which less prominent scientists rejected outright. Pauli himself postulated that it couldn't reasonably be explained in any other way because of the circumstances and frequency of occurrences. Many other scientists who witnessed these events concurred. And in fact in April 1948 on the foundation of the C.G. Jung institute the Pauli Effect once again occurred. A Chinese water vase without cause fell to the floor causing a flood upon Pauli's entering the room. This on many levels aside from the obvious perceived telekinesis created multiples of synchronicities in a ripple effect. These ripples to me meant that these perceptions should be given care to know when enough would be enough? So as not to over read the message. What to most would just seem like a single unfortunate occurrence was interpreted as an alignment of meaningful subconscious messages.

Remember, many people say: "There is no such thing as a coincidence!" I'm not one of those people who believes that every coincidence has meaning. A coincidence is sometimes just a coincidence knowing which is which, is the trick, besides knowing the message.

Pauli who was one of the pioneers of early quantum physics along with many others believed that beneath the subatomic layers there are still other unknowable hidden layers and dimensions. The dilemma once again was that it could not be proved scientifically yet. His world of science had no references for that which could not be tested by scientific means. That's too bad because in parapsychology almost nothing can be proven. Still he was part of a growing group of scientists who are convinced there are hidden unknown and as yet unprovable deeper dimensions. The number of which is a guess with some saying up to twelve. And with Bohm saying they could be infinite.

Together Jung and Pauli helped pioneer the study of parapsychology by introducing real science and controlled data gathering techniques. Together they came to the same conclusion from different fields of study but with Jung working through the psychic dimension and Pauli the physical it was a match that seemed pre-ordained.

Now though, others studying non-material or even fringe science also advanced the field (sometimes unknowingly) by the very nature of their work. Pauli's own early work in quantum physics had a strong influence by changing how we think about the behavior and properties of matter itself on the quantum level. Also what may be one of the most important unintended discoveries in the study of the paranormal is Einstein's quantum entanglement although he himself had many misgivings about it. Many today believe it may hold the secrets to understanding synchronicity, ESP, telekinesis and a number of other psi abilities. The "connection" between mind, matter and consciousness bridged by entanglement through multi-dimensions seems like a very promising avenue of exploration.

Entanglement as it relates to all paranormal and its seeming ability to completely bypass the spacetime dimension for one, while also working directly and simultaneously with and through all the other known dimensions including consciousness. This is amazing stuff people. Based on real science and although unaccepted most of it is put forward by some pretty accomplished physics and philosophical legends.

A great solid example of good philosophical theory is Rupert Sheldrake's "morphic fields" in which he shows us how fields create relationships across both time and distance. Physicist David Bohm's "Implicate and

Explicate Order" as I previously mentioned. Michael Talbot's famous "Holographic Universe" or Physicist David Peat's "Meaning and Form." All of these once again propose a hidden connection beneath our material and temporal reality. What this means is that much of our true reality is hidden from not only all of our senses, but from our cognitive ability to understand all that is going on. Just as in Plato's "Allegory of the Cave," we can't know the world as it really is with our limited sensory information. Only in this case of pre-existence it's because it doesn't yet exist.

Jung gives us the example of archetypes that in turn influenced Sheldrake's thought on morphic fields and the lattice that underlie the crystallization creation process. It has no material existence and yet it gives crystals uniformity. We can only know of it by the resulting crystal's properties. The uniformity of implies an invisible design or blueprint. Sheldrake would say in his theory of morphic resonance: "It is because it was before." Meaning it gets its form because the crystals that came before had the same properties. Jung uses this as proof of a hidden dimension of "archetypes." This is why synchronicity, the crystal or snowflakes take the form they do. Scientifically this is impossible to prove because there is no empirical data to support it when there are no physical designs. We know only because we see and then reason with our own minds. If you've experienced seeing a ghost or synchronicity or any other form of the paranormal the only acceptable proof you can have are your experiences.

Still we can all take comfort in knowing that as Emmanuel Kant said "no one can know the world as it really is." Even the top physicists in the world can only guess. Much of quantum physics is competing theory and it's the same with non-material science. What we must keep in mind, is that all of quantum theory is just that, theory - dozens of theories. I myself have read over a dozen so if anything only one can be right. I myself have read many scientists insist that many of my favorite science theories are wrong or made up. I myself have also witnessed much of the impossible with my own eyes.

Then there's the problem with consciousness. Consciousness itself cannot be quantified - there is no explaining it. Many believe it exists outside the brain, perhaps the Akashic field. Psi-fields, source fields, the flow, these are all conceptual dimensions existing outside of our frequency of space-time bound together by abstract relationships. Where so often consciousness and matter are believed to be two aspects of the same thing (unus mundus). These are the domains of the trickster gods where Hermes and Loki conspire to fool us using mind, matter and meaning.

Ok, and although I'm in agreement with many of these theorists about some hypothetical matrix. **What is the intelligence behind the curtain?**

Since before recorded history humankind has recognized the existence of a greater intelligence. Ask any believer of any religion or any other esoteric system and you'll get many different answers. The spirit, God, the Goddess, Sophie, Allah, the Demiurge, The One. History's greatest scientists came to this same conclusion. Einstein said: *"Everyone who is seriously involved in the pursuit of science becomes convinced that a spirit is manifest in the laws of the Universe - a spirit vastly superior to that of man, and one in the face of which we with our modest powers must feel humble."*

Also Max Planck the father of quantum physics said: *"All matter originates and exists, only by virtue of a force. We must assume behind this force is the existence of a conscious and intelligent mind. This mind is the matrix of all matter."* Isaac Newton believed the universe was mechanical – like a precision clock set into motion by God and then left to run. There are others who believe that all existence is an emanation of God. Some don't believe there is an external intelligence at all. This is not my belief!

I have a unique prospective on whether or not there is an external intelligence without the need for deduction. Einstein, Planck, Jung, Pauli, Peat, Bohm all knew there were unseen, un-knowables through their own experiences. Many others like philosophers and theologians also believed through mostly faith. I know that there are more than just hidden dimensions or mystical hidden intelligences. This doesn't mean that these words are my attempts at making an argument for the existence of God. Only to say that the vast majority of the planet believe in the supernatural. Gods, angels, spirits, witches, demons, the Devil, magic, voodoo, shamanism, these are all supernatural symbols or archetypes of human believers in something beyond or greater than ourselves and I am one.

The coming together of both mainstream science and religion is gradually happening due to the research started long ago in quantum physics and perhaps psychoenergetics with the inclusion of consciousness.

Synchronicity and the Secret of the One

What many of the ancient esoteric belief systems and modern science have in common, is the belief that your own thoughts can alter the outside world in relation to you. Though, you alone are not the creator within the flow of the source field. You are simply part of the process. The biggest secret of synchronicity is not that there is a co-creator, but that it's YOU! Ultimately there is a separate coordinating intelligence in control.

This is what causes events around you to coincide without your fore-thoughts, like for instance, the moment of my birth: 3333. That's why we pray or chant, recite mantras, sing praises and invoke - expecting a particular outcome. Deep down inside we've always known we were being watched, even in an empty room, we are never really alone. How many times do events align that seem so strange and statistically unlikely to have occurred by chance? Without our thoughts, these synchronicities must originate externally. This means that "The Source" or "The-One" controlling it all is out there and not within… but then the bigger mystery is: "what is meant by the one?"

"Synchronicity is God's way of remaining anonymous."
---   Albert   Einstein

Chapter 4
The Dilemma: Knowledge – Belief

Well what do you believe - do you believe in anything? I grew up in a small shoreline New England town of around twenty five thousand. I went to a small Christian church every Sunday with my family where we read and listened to bible stories. God, Jesus, Moses, the Ten Commandments and also much interpretation from those who were sermonizing. I do remember being intimidated by the power of the great mightiness of his most holy as we were young. Only, I doubted the reality of any of it and not that "I knew." I did learn to pray though and often just making up my own just in case and even when I was alone or in those desperate moments that we all have. But is there really a need to have any belief system at all? Do we as humans need to believe in anything for those times and is that a weakness or is it a way to give us strength?

Is belief a weakness or a need, or doubt in a solid commitment to some empiricism? Well I myself told you that I believe that who or whatever is controlling everything is external to us but having this external belief and not knowing who or what it is, creates a dilemma. This is actually my biggest dilemma with belief because

I've experienced all different sort of paranormal or supernatural phenomena that I have no solid proof of what it is that's causing this to occur. All I know is that it's coming from somewhere else and for me, it's like there's an entity observing just outside of my reach - watching. Like that day the light appeared. Science can tell me it was magical thinking but it is not a synchronistic event that could pass for imaginary coincidence or that could be mistaken as a coincidence. It was as real as anything else I know to be real and I believe it to be a cause. I long ago accepted that it was an entity and believed it was there for me. Since it was the second of three separate encounters that I had. And so wrote about it and this led me to one of my biggest synchronicities that I had. But this is how things went for me as far as compound macro-synchronistic events.

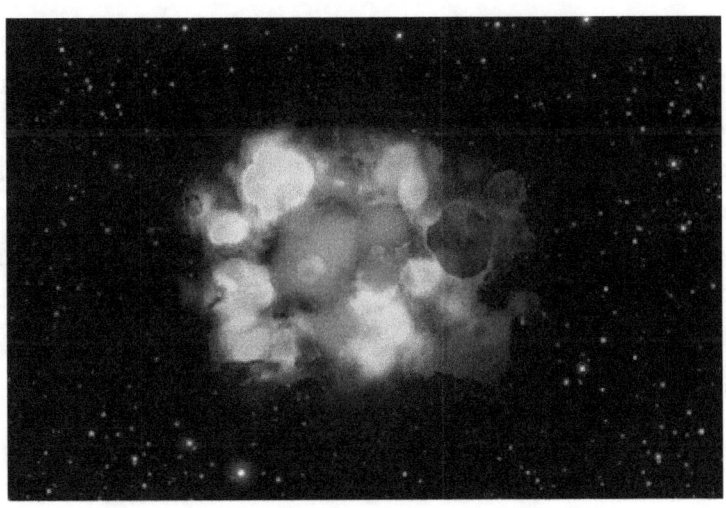

Above: My artist rendering based on actual public domain UFO photo altered with Photoshop to resemble what I saw. It's almost exact but filled the sky a mile wide above me.

First of all in retrospect the entity impressed me as God-like. It was an enormous multi-colored luminous cloud like object about a mile in width. It was a globular, slightly oval luminous plasma with about a dozen distinct but separate colors. Each color was also globular, changing shape and size while adjacent colors would either contract or expand in response. The colors were bright, saturated and fluorescent in appearance like neon. The overall size never varied, it was completely silent, almost directly above me and I seemed perfectly centered from left to right, beneath it.

After about 30 seconds, the thing just shot straight back and disappeared into a clear sky full of stars. Then after about ten seconds, it just as suddenly came back to the same exact position and size with all the colors in their exact same places. The speed was beyond belief, taking about a second to travel the distance. WOW! From a mile across to hundreds or even thousands of miles away in one second. It took no more than a second to return too. I remember thinking: "This is weird and I don't think it's from here" but here it was.

This time it hovered for ten seconds both beautiful and silent. Then once again it shot away for another couple seconds before returning for the third time. Finally it went back and forth three more times within a five second span as if showing off its speed and then it was gone. I stood there frozen another thirty seconds before suddenly being jolted by the realization of what I had just witnessed. I just started running faster than I'd ever run before. When I got home heart pounding, out of breath, I frantically told my older brother Frank, "I just saw a UFO!"

The next day, my brother told me that he read in the paper that other people reported it and that the explanation that was given was that it was gases escaped from a test rocket. Does that sound familiar? That cover-story never sat right with me. No rocket testing that I ever heard of was done over populated New England towns. Anyway, since when does escaped gas take off, come back, take off again and come back three more times before actually escaping?

This was written and would later be posted to many sites and used for magazine articles and while I was writing about my encounter a storm "Irene" hit. On that exact beach over thirty houses were destroyed with some being washed away but no other houses on any other beach in town. Those houses were there over a hundred years and no storm had ever occurred like that before in the history of CT. It seems strange that it happens during my lifetime while I'm writing about that exact beach?

I remember thinking back to the time that only a creator type entity can generate the pre-existence leading up to created scenarios of the multiple synchronicities I've been experiencing. I can imagine that only an entity having god-like powers would have these capabilities.

Now this story was quickly shared by many and the one thing that struck me is how many religious sites shared it. Christian church sites and other religions with people suggesting it was an angel. I didn't respond but I began to realize how many people from all over the world believe in a god, angels or other supernatural beings. But they relate them to what they know. Not everyone knows science, but almost everyone relates to a deity.

Then there are all those who say our own minds are working in unison with the universe causing it all subconsciously – although I have my doubts. I suspect that knowingly or not that much of this is directly through paranormal abilities for many of us. But not this, this was not that. Also for me this initiated a personal transcendence. From what I've experienced and from all I can tell, my scientific self says: "A Transcendent Alternate Existence" or to an alternate reality. Not from here to another place but rather like opening a window or doorway in your mind to a realization that there is much more you don't know. That there was something about me. Like that I am the cause or you are the cause. Like the universe is trying to tell you something. This was Wolfgang Pauli's realization with psychokinesis or Jung's realization when he broke Freud's shelf which I'll get to - they were experiencing the anomalies which made them aware.

Now I'm not saying that any of us who experience the paranormal are all the same, but if you're anything like me, then that knowledge wasn't derived from reading it in a book because much of it seems to come from our minds themselves. It's from experience - "knowing" and it's easy for me to say: "I just know" because of a feeling but my knowing comes to me from an external source. Not just because I lived it and do live it, but understanding its external origination because of it. Like my thoughts are materializing with or without a medium. I know it's not very scientific right? But I know because I was "shown" from real events which led me to this place that's not very scientific either right? I only recently realized there was an entire language for. It was through reading many others who also knew from their own personal experiences. And they also may have been academic types but they saw as I did that something is wrong with the accepted science. They also looked for answers outside the rigid faux box.

Belief however, often has a double edge because it doesn't have to be true. We only have to decide that we believe it's true. It's one of the strongest emotions any of us will ever know. We have to have it in order to get most things accomplished or even started. That and certainty because that same double edge can often be misinterpreted. Many great human endeavors are initiated based on optimistic outcomes. And so how we interpret things may mean everything. Usually this is based on intuition but all of us have our own methods of intuiting. A priest uses the bible based on the belief in the words of God, a scientist uses information he or she believes is true empirical data. The fortune teller relies on the tarot cards or a crystal ball and seers perhaps on the stars. Yet no matter what method you use to interpret, it will take some wisdom, some type of learned or experienced knowledge and maybe even talent.

For me that answer is simple because I didn't have to try. I've seen and experienced so much paranormal that I could believe in almost anything and do. After the entity that came down from the time of my first two encounters I was never the same and believe that almost anything could happen. Also in my case I feel that I am grounded by all the actual research I did and still do in order to somewhat justify everything I've seen. Reading the stories of so many accomplished and distinguished others may have saved me from believing I was going crazy. Like one of my favorite old time classic synchronicity stories about Jung and Sigmund Freud. Most refer to as "the exploding bookcase" involving what Jung called a "catalytic exteriorization phenomena" which was actually in a movie about them. The movie is called "A Dangerous Method" about Freud and Jung's tumultuous friendship. Now you should know Jung and Freud had a whole working history in the backstory, so you might want to look it up. But I'll paraphrase here from several different sources but: According to Jung – Freud was alarmed by one of his letters. Jung's interest in synchronicity and the paranormal rankled the strict philosophical materialist: he condemned Jung for wallowing in what he called "the black tide of the mud of occultism."

FREUD   JUNG

Just two years earlier, during a visit Freud in Vienna, Jung had attempted to defend his beliefs and sparked a heated debate. This caused Freud to dismiss Jung's paranormal leanings, "in terms of so shallow a positivism." Suddenly a shocking synchronistic event followed. Jung writes in his memoirs: While Freud was going on this way, I had curios sensation. It was as if my diaphragm were made of iron and were becoming red-hot--- a glowing vault. And at that moment there was such a loud report in the bookcase, which stood right next to us, that we both started up in alarm, fearing the thing was going to topple over on us. I said to Freud: "There, that is an example of a so-called catalytic exteriorization phenomena." I now predict that in a moment there will be another such loud report! "Sure enough, no sooner did he say the words that the same detonation went off in the bookcase. To this day I do not know what gave me this certainty. But I knew beyond all doubt that the report would come again. Freud only stared aghast at him. Jung told Freud that this is a perfect example of paranormal phenomena.

As this little anecdote shows you, Jung believed his subconscious mind directly affected the real or physical world because of the tension for whatever reason. There are many personal interpretations that were involved like rejection of the father figure which Freud had become to Jung at the time and so on. These things are important to me as well. I look to see when signs align for the meaning (although they are often not that obvious) to help guide me to what I believe will be a good path for the best outcome. In this case it is something that's required for anything having to do with esoteric sciences or magic? Do you believe in magic? The strange, the mysterious, the paranormal, God, angels, demons, things you can't often see but must know. At this point the normal or real three dimensional material world has been set aside once you've accepted the belief in things that must be intuited or experienced.

As you can also see this type of thinking also often meets with rejection from people who have their own beliefs and in this case expectations of one another.

So what was my interpretation of the entity? I only said I didn't know what it was and could only classify it as a UFO. I only later thought it was an entity because of the synchronicities that occurred years later when I was writing about it. If nothing happened at all at that exact time I wouldn't have made anything more out of it. But also because I didn't believe in anything when this all happened. I was a regular person who didn't believe in ghosts, God, or the paranormal. Then this light shook my faith! If this could exist then why not anything else be possible? What if I was wrong about some things? Ever since these other events that kept happening I have to believe these were big messages and they were meant for me - everything was meant for me.

Chapter 5
The Science Dilemma

This openness to belief brings about another very big dilemma for me that I just recently realized I had and that's how quickly science is changing. I wrote an article about a year ago for a magazine in NYC called The Epoch Times and many sites online. It was shared very quickly around the world - many countries, many languages... The one thing I realized but was unknown to me as I was writing was that my research was too cutting edge for some people. In fact it was unknown to most readers like even editors – because the science was so new everyone required citations. So I knew I would have to explain the science within the article for the readers who aren't going to read the citations. It was called "Is Communication from the Future Already Here?" My belief of course is that it is! Computers from the future are already communicating back through time to other computers. And because of this who knows what unknown other communication or advances have been going on? In a nutshell this is done using the same theoretical science as I believe justifies most other paranormal science. My belief is based on subatomic particle entanglement physics progressing and my own speculation.

Synchronicity and the Secret of the One

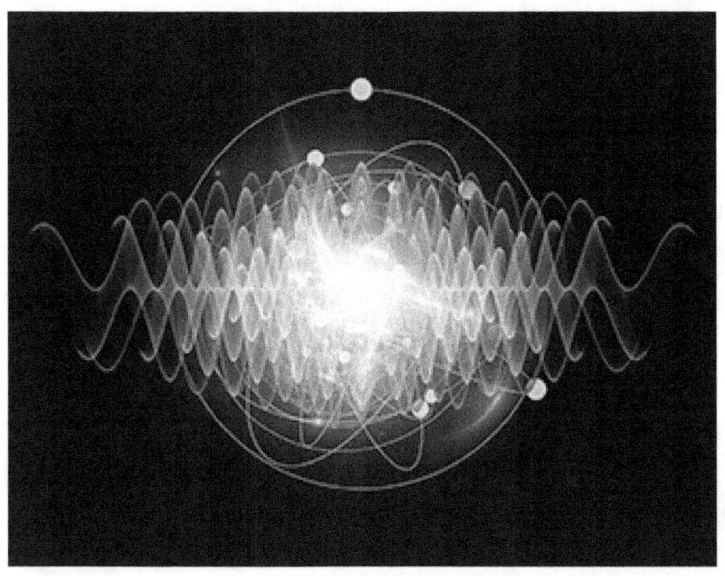

My belief that "entanglement" breaks the time barrier is now very newly being established science and was even newer when I wrote my article. This was based on collected current science but when other science freaks like me read this I was given the third degree from publishers to editors to skeptics. It seemed that almost no one knew that a theory of entanglement was that the photon or electron entanglement communicate faster than the speed of light and at any distance instantly by seemingly bypassing the spacetime dimension completely but they also seem to go backward in time. Yes, meaning they can communicate backward to an entangled particle before the measurements were ever taken. I know it seems impossible but that's why I LOVE science as much as I do.

I don't think there was one editor that didn't require scientific proof and citations of what I was saying before publishing. The science editor for The Epoch Times went back and forth with me for three weeks wanting more proof each time and based on our discussions she read everything until she grasped what I was saying enough that she would publish it. They have large circulation worldwide based out of NYC so they don't just print whatever someone says. To my surprise none of the large websites did either without my proving my own hypothesis. That's what I had to tell people, that this was my own hypothesis based on solid science.

One of all of our problems when it comes to science is it's not only that it's moving so fast on the theories but they're being pushed out one after another. Because of our hunger for hot new science it's not meeting rigorous standards of verification before making all the science sites and publications. That's our fault mostly because we want words and there are so many publications - we need material. But most of the established classic quantum physics that is the foundation is decades old. One of the most interesting is the newest challenge to the "loophole-free" Bell's Theorem which is from 2015. This means more years of what should be scrutiny. Which means more testing and more years to move the bedrock. But the original theorem was published in 1964 and just as recently it is being checked and rechecked in

2015 for a possible update. If I had said this a few years back everyone would have been skeptical. And now seeing how skeptical people are, what about me? I mean while I'm doing my own fact checking am I careful enough not to let the ideas of others lead me astray? I after-all have to put my stamp on all that I put out there. With so much theory we should be careful of what we believe. It's 2018 and the loophole free challenge hasn't been verified last I checked.

For me one of the other biggest problems is that you want to also educate your readers if you can while at the same time making it readable. I call this popular science and it's where I differ from a lot of my fellow science writers. Trying to make my articles understandable and more accessible takes more time but if you read my article "Communication from the Future" or others, as a science writer it allowed me to reach millions of readers rather than thousands. The numbers to me aren't as important as the dissemination of knowledge or ideas in a positive way. Hence – pop science.

Well, but even though, I do the best I can by fact checking in triplicate because I don't want to get what I do write wrong. Even when I heard a well-known quote from Einstein I checked three sources. Guess what? All three sources from well-known and reliable publications were worded slightly different and after checking I read that Einstein had been translated from German and that others noticed the problem too.

Even more and beyond that there is the problem of science that most writing about science don't have to be concerned with. That is of course what I've mentioned from the beginning - esotericism. We are using a science on a subject not accepted by most - an alternative science or psuedo-science with alternative rules, alternative physics and I don't even know that you can call it physics. If our known physics laws are the bottom line when it comes to our understanding of our world then it does require a new science. Remember physics deals with matter and everything is made of matter even the smallest subatomic particles. And they obey the laws of physics, but even quantum physics don't obey the laws of Newtonian physics. As far as we are taught with our science. But Tesla says:

"The day science begins to study non-physical phenomena, it will make more progress in one decade than in all the previous centuries of existence." --- *Nikola Tesla.*

In the context of what concerns us what did he mean? For the answer - I go back to a previous Tesla quote:

"Today's scientists have substituted mathematics for experiments, and they wander off through equation after

equation, and eventually build a structure which has no relation to reality."

Tesla – believed much of theoretical physics lacked observed reality. Many times humans see things that are not swamp gas or morning mist, or coincidences.

What we are dealing with here is something else. Something that effects matter and we don't know how. The paranormal, witchcraft, magic, the supernatural and it may all be involved. When it comes to synchronicity it may just need to get a little deeper because we are dealing directly with the mind, the universe, consciousness and often without our intention. According to Jung, Pauli, Tesla, etcetera our thoughts and often subconscious are directly affecting the outside world along with our inside life past, present and even future so this is why we need to understand. We just see it occurring while science tells us we're wrong and that's why I had to see for myself and that's why I gravitated to the research something that has only very few facts. That's what led me to seek out the work of the scientists I did. Many of whom were accomplished in their own fields but were looking for answers just like me or any of us. These aren't just scientists and researchers, these are icons of their fields.

The facts we do know are: "Don't expect this to change anytime soon" because this would mean a change in the paradigm and if you are looking for change that doesn't get research funding. Aside from that, do you really think that the powers that be want the average man or woman to know about the study of the existence of an alternative science yet - or about fluctuations in time and space that allow people to see ghosts or the future or the past as it relates to the paranormal? Starting with the executions of witches, sorcerers, alchemists, magicians up to today, they have put too much effort into keeping hidden and un-teaching us about what used to be accepted knowledge.

In 1957 Dr. Heinz Eyncnck (University of London) wrote a letter stating that there was a serious conspiracy involving 30 major universities around the world and hundreds of respected scientists to silence the truth about what we knew about the powers of human consciousness.

In fact, and but not to get too technical since this book is supposed to be intelligible to anyone, it is believed that going faster than light (superluminal) also violates the law of causality. This type of research is exactly what we need to help explain some supernatural phenomena like ESP and telekinesis. However many scientists like Tesla suggested that we need a new science model for a new era. Science which includes human consciousness as a new mathematically represented constant to become part of the equation. Below is the "*Psychoenergetic* Science" model, From Dr. William Tiller's *White Paper VIII* introduces *"consciousness" as a new constant.*

Mass ⇔ Energy ⇔ Information ⇔ Consciousness

Yes I get the irony! On one side trying to solve problems and on the other we identify that some are trying to hide the solutions. Dr. Heinz, Charles Fort and our good Dr. William Tiller who had to find private funding for his research were all aware of that. What is it about consciousness that these institutions of higher learning would conspire to keep us from finding out about?

Chapter 6
The Philosophy Dilemma

For me philosophy is like science it's not only all over the place but it's really only offered as systems of belief much like religions and in fact including religion. What is philosophy anyway? From what I've read and how I understand it, it's something that most writers give up on trying to define and in my opinion because it's a heady subject both subjective and confusing. It's in the eye or mind of the observer and it is just or could be your own opinions on existence itself. And that my friends is why we might defer to what's called Cartesian philosophy or philosophy having to do with Rene Descartes. And it's a matter of preference but to me it's the most succinct explanation of philosophy.

When we are discussing any of this and we admit that it's all at the unknown level then we have to admit that the more we read the more we realize how little we know. Yeah, just like quantum physics except there will always be someone who claims to know more. So how can we say there is an expert in any of these esoteric fields of study? This all begins with philosophy thousands of years back and thousands of years later you see too it's all intertwined and many may say just as senseless. Back in ancient times the Egyptians, Babylonians, the Greeks - magic and philosophy were part of science and medicine and they prayed to gods with many gods for every different affliction or ill. To make it rain or to have a good harvest. Just like much science it's all different beliefs and just like science we go from the subatomic, to the mind, to religion, to pre-existence and how we think about any of these areas is all based on how we perceive them.

That would be based on what teachings we choose to follow. The words we use have meanings and with philosophy much of the meaning only the original writer can understand. I was once trying to paraphrase some amazing ideas from George Wilhelm Friedrich Hegel but his writing is difficult to understand at times. Also his thinking is circular and as Charles Fort would put it, *"a circle leads to nowhere."* His contemporaries also complained that his writing was too difficult to understand. So even though his contributions to philosophy were very influential throughout the world, other philosophers said he didn't make any sense. After much of my own reading, I have to agree that he is very difficult to follow but it isn't just Hegel, much of philosophical writing is.

Many critics also suggest that philosophy is useless except maybe to metaphysics which also carries the label of useless but personally I see it as a powerful tool for both political and social science to point out just some usefulness. Also to "ontology" and "epistemology" or the study and understanding of knowledge and existence itself. Now I'm not going to overwhelm you with much of these philosophical arguments on the nature of reality, because for the intellect as much as it is an eye opener it carries a heavy learning curve it can also be very confusing – especially for our topic here. I'll touch on a couple I suppose like so you see what I mean. And what would a discussion on philosophy be if I didn't mention the father of Western Philosophy himself Rene Descarte? He is best known for his philosophical statement "I think therefore If am" in 1637. I think therefore I am what and what does he mean by it? The fact is that we can't prove anything is real according to most philosophy but that we even think at all was sufficient to prove that we are at least self-aware so we are beings of thought.

Two of the main divisions in philosophy are mind and matter in other words we exist or for that matter anything including us, or we don't. Things exist or they don't and this is called either realism where existence is real or idealism is, where things exist only in the mind. Everything including ourselves in which case it is very easy to imagine synchronicity because it's all in the mind. No actual physical humans ever had to cross paths for the coincidences to occur or the psychokinesis of Pauli, or the wood in the shelf to break because it's all just mind stuff. Just like the statement in the previous paragraph no one could really prove that something is out there in a material world. Like "Descartes Demon" in which there is an evil demon tricking us into believing we have bodies and there is a world outside of our minds. As Descartes has written, the only thing any of us can really prove is that we can think. Other than that, we might only exist as minds living a dream and even other people are part of that dream. We can after all only prove that we have the ability to think. I can't look at you and prove that you have that same ability to think or that you aren't part of my dream.

Rene Descarte was labeled a "rationalist" by the way, along with many others like Spinoza. Many others had many other beliefs like "empiricism" similar to rationalism meaning that knowledge comes from empirical evidence. Then of course "realism" and Plato. "Idealism" and probably most having to do with justifying philosophy in our discussion of synchronicity - it's the mind stuff. And isn't that why we're here - the mind stuff, the science, the magic, the supernatural? The "RECIPROCITY!" Maybe that has something to do with it? A give and take with the universe… Maybe that by far is why the universe gives us these messages. Mind and matter either or both, the combination – belief. And leave everyone to decipher the massages. That my friend is a no win game the battle of philosophies.

Before I go or go on forever about philosophical belief systems I wouldn't be doing any of us justice without mentioning first Aristotle who was an ancient philosopher. A long time student of Plato and a scientist in both physics and metaphysics. The fact is that this isn't about philosophy but you should know that everything seems related and before science these ideas of reality were around - thousands of years before! There are really a lot of books dedicated to philosophy that could give you much more justice than a book about synchronicity. After all when we involve Kant's idealism everything that could be real isn't.

Pythagoras was said to be the greatest philosopher and came up with advances in math (Pythagorean Theorem) and music formulas and actually believed everything was mathematical including the movement of the planets. He said: "all things are numbers."

St Augustine another great philosopher who has written extensively and was also very influential and mixed religion and philosophy along with St Thomas Aquinas who believes everything is God. To him you need look no further for the "one" since we are all emanations of God. Between the two of them there is a lot of lengthy written debate concerning "determinism vs freewill."

The fact is we could go on with philosophical ideas forever and right into another book. You literally could get lost in philosophy because of the many beliefs merging into one another. Taking parts and rejecting what seem out of place pieces.

Chapter 7
Dimensions of Consciousness and Meaning

How many dimensions are there? So if you go along with the consensus, the average is twelve and with some saying infinite. Infinite dimensions with others suggesting infinite universes. And we all begin with the three here in our known universe - width, height and depth and include the fourth time. Then we add the forces like gravity and energy and much of the rest becomes entangled in personal belief. All that I've looked at include the creator entity of a god if you will and then depending on belief, which god? Even the heavy science theories as opposed to the spiritual ones include a creator. Bohm, Einstein, Schrodinger, Bacon, Galileo and many others believed in a creator often the creator is the first dimension. Like God is often the first dimension when the creator is God. Since whoever the creator is, he or she created the universe from nothing, outside of what we recognize as existence that is somewhat a given. But in that there is no consensus as to the order or as to what the dimensions are. The first four are usually the same but what they are after that and what they consist of aren't. Energy, gravity, consciousness - consensus? I'm still thinking ancient gods.

Synchronicity is actually unfolded in one of our dimensions of meaning as our reality is and it is heavily dependent on our consciousness in order to have any meaning at all. The universe actually requires consciousness to have any meaning. If there were only an empty galaxy who would it have meaning to? In fact I've mentioned already that Tiller along with other scientists believe that consciousness itself should be a constant. Part of our theory on the nature of reality. Since the beginning of quantum physics research it was apparent that the minds of the researcher were altering the experiments. Eventually some thought to just include them as part of the equation. Like mass or energy, some even consider thought to be a dimension like spirituality, even an original concept of the flow like Bohm puts it out there. At times he was questioned about this because to some he seemed to be dabbling in the mystical with his language.

However, on different occasions some scientists while engaged in their normal straight forward research, were seeing anomalies within their experiments. Thinking perhaps they were errors, they repeated the experiments over and over and again "the anomalies." It was obvious that the minds of the researchers was exerting some hidden force that was causing these anomalies but it was not measurable like gravity or electricity were. Some physicists who perhaps naively went into physics to study textbook physics, wanted to get to the bottom of the anomalies. For some it led them into switching their research over to researching the anomalies. Of course for others the interest in researching parapsychology was their original intent. Perhaps even knowing about the anomalies in advance.

Now I should add that Dr. Tiller is just one of many doing research into "Quantum Consciousness" and or the power of intention. Stuart Hameroff, University of Arizona, physicist Roger Penrose, Cambridge/Oxford among others have published much material written on this subject. There are many competing theories as far as Quantum Consciousness goes. They all agree that Newtonian physics cannot account for consciousness, which separates them from the mainstream. Simply put, that consciousness is not a function of mechanics of the physical brain as most believe. However, how consciousness operates is all still theory. The actual mechanics of the brain is still very mysterious. Many also believe quantum computing is a stretch.

Without getting too technical Penrose and Hameroff theories suggest that the brain is doing quantum computing and many others suggests that consciousness resides outside the physical brain, in the universe (universal consciousness) and that our brains simply receive the signals (thoughts) like a radio frequency receiver. Another theory is that all matter has consciousness and that we simply decode it. One of my favorites is we aren't really here at all and really we are just part of the conscious universe giving us the illusion that we have bodies and that there is matter around us. It's sort of like we're characters in a computer game. Sounds much more like an update Emmanuelle Kant's idealism.

This leads me to back to the word "noumenon" which is an interesting word in philosophy and metaphysics which is also a heady area of study. Almost everything having to do with synchronicity or all of this type of esoteric science is strange. Noumenon is something you can't perceive with your physical senses. It is known by deduction and intuition just like many of these things we experience when we have scenarios of involved synchronicities. Since metaphysics to many is an unacceptable science and philosophy is often too difficult to understand. And although you might be taught different because most of this is all mind stuff. Philosophy to me is just as important as, if not more so than. Especially when much of our science and these concepts seem much more useful to the sorcerer or the fortune teller.

Much of our social science is based on philosophical ideas and differences. Our philosophies have led to the building blocks for the foundations of our civilizations. As far as our esoteric beliefs they are also the building blocks of our sciences because they are partial causes of our formations of our own beliefs and strange ideas outside of science. Many of us believe what we believe against the facts that we are told it can't be so. The means we use to justify whatever it is within ourselves or what we choose to believe.

It was only through accomplished thinkers that have personally gone through much of what I have that I found truth. Much of what I myself have written in articles has been met with some type of criticism. Often times people will question my science without realizing that no parapsychology is provable and that it's almost always witness testimony else shamanism or wizards.

Prove you saw a ghost, you have ESP, saw a UFO. Even with the best clearest photos in the world and with ten total strangers that say they saw the same thing, you will be challenged to no end.

Even modern philosophy is skeptical of the possibility that there is noumenon being independent of our senses.

Chapter 8
"The Flow"

*"Space is not empty. It is full, a plenum as opposed to a vacuum, and is the ground for the existence of everything, including ourselves. The universe is not separate from this cosmic sea of **energy**."* – David Bohm.

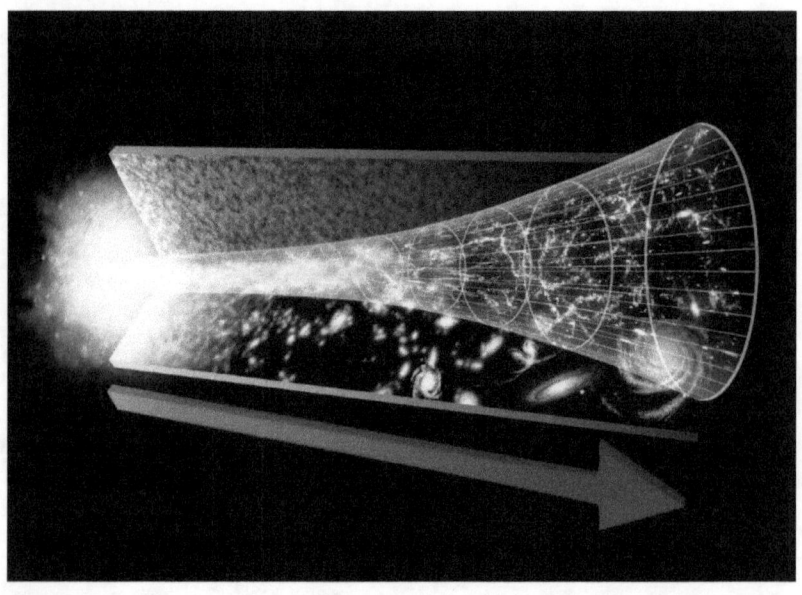

So what is the flow? In the physics of Bohm, this is the flow the flow is the entire universe that he called the holomovement and every hidden dimension and every subatomic particle and the whole of everything you see and everything you don't see. And I mean that literally. It is the undivided entirety and the never ending state of change moving as one, folding over and over in the process of creation. It is the implicate and the explicate order. The implicate order is everything endlessly folding into itself and then unfolding to create everything in existence. Even us. It is a flow of information in any belief system. It's evolution - everything is on its way to becoming something else. The evolving sea of creation from which everything springs forth and this is similar to how Bohm who created the theory describes it. The explicate order is a small part of the implicate order and we call this our three dimensional world. He talks of all of man being part of the whole without separateness and it all sounds very modern mystical. And supposedly we can't detect the implicate order.

Ok so then how do we know it exists? We know because we have accepted that the flow is part of the anatomy of universe as a whole and that together we combine as participatory features of it. I even suspect that many of us didn't even need to read or didn't even know there was any science behind it – which I am one. I believe that much of it is still manifested in or through mysticism. Again combined we manifest existence from the seemingly nothingness of pre-existence. This is only because of our belief or trust in personal observations. We can see that much of the strangeness is personalized to our own thoughts and much like we helped to create it right from the edges of our consciousness or dreams with our own personal conjurer. Most of that is also because of our belief and trust in the observations of ourselves and others who came before us and have helped to create an identifiable language we can relate to for us. So we can make that choice to believe. Or - we can accept the argument of skeptics that it's all only proof of inevitable randomness, or once again "Magical Thinking" or we are crazy. Another choice is to use the highly accepted and most often more true proof of "observation." All our top notch theoretical physicists of all time can only hypothesize as to what is going on. And as strange as it sounds "The Flow" which sounds a little new age is a very common term that's used to describe what they came up with.

"The Flow" as we are using the term here is a description for pre-existence which was actually given the name the "implicate order" by theoretical physicist David Bohm in the early 1980s to describe a non-existent non-material reality which includes folded space and time and everything else. If you are confused here, it's OK because none of this definition existed before quantum mechanics was officially born and named by physicist Max Born in the early mid 1920s. Since then it's all hypothesized by non-provable theoretical physics mostly. I mean if you want to create a science for magic, then who better to do it than scientists with physics degrees? Anyway, it is far too conceptual for us here to do it justice because it is all still hypothesized concept not known by anyone. The saving grace for me is that the experts that I usually quote claim to have had paranormal experiences just like I do.

Speaking of grace let me just mention there is another pre-existence which we are not speaking of and that has to do with the soul and different theories about how it came to be. We are supposedly non-existent till the soul enters our body and there are different theories how that happens as well. We are born with one - it enters our body after we are born – after we are baptized, and so on. But we are not speaking of the soul in this case.

When I first began to consider synchronicity and what it meant I immediately contemplated the flow as a concept of streaming thought flowing down to a river metaphorically of course. Really I only knew it as being of a more spiritual and paranormal form of thought of it in that context. But that's what I think most people think I don't think most people look at it and think science. I mean like everything came to me from the point of view from seeing the light maybe the supernatural spiritual world and not the science. Everything had a different meaning to me then - it meant something was reaching out to contact me and it was a mystery. Something was trying to get my attention – to tell me something. I guess that even though the science is not giving me the answers it's filled me with many questions that could be the answer and for others. There is a hope that someday maybe it will. See I do believe in magic and all the supernatural things that I see as real so I might be ruined. Now here trying to tell you I have to revert back to others to rely on their many years of wisdom.

Because my opinion is that everything is always coming from somewhere else and so that part it seems we have all reached the same conclusion. That is that there is a super intelligence running the show. I'll paraphrase some ideas from others here. "Bohm believes that there is an underlying cosmic intelligence that supplies the information (the player) like a computer game who is the third category. Following this analogy, Bohm sees the whole process as a closed loop; it goes from the screen to the computer to the Player and back to the screen.

Bohm's theory of the Implicate Order stresses that the cosmos is in a state of process. Bohm's cosmos is a "feedback" universe that continuously recycles forward into a greater mode of being and consciousness always processing, always updating with new data that it gathers and then incorporates.

Bohm believes in a special cosmic interiority. It – "the Implicate Order" and it implies enfoldment into everything. Everything that is and will be in this cosmos is enfolded within the Implicate Order. There is a special cosmic movement that carries forth the process of enfoldment and unfoldment (into the explicate order). This process of cosmic movement, in endless feedback cycles, creates an infinite variety of manifest forms and mentality. Bohm is of the opinion that a fundamental Cosmic Intelligence is the "player" in this process. It is engaged in endless experimentation and creativity. This Player, the Cosmic Mind, is moving cyclically onward and onward accruing an infinity of experienced being!"

Of course there are others who believe that the universe itself is alive and conscious and that would explain a lot but it would have to be a super-consciousness like Bohm says. If you can imagine how many stars galaxies and how many universes there are and for how long? How many species of all types of living things that are sentient beings so I begin to think - then why would synchronicities be created just for me? Not just you, but because you are part of the universe and they may just be a reflection of your thoughts as part of that living universe. That's only guessing out loud though based on Bohm's feedback loop!

Remember Einstein and Max Plank both believed that we are dealing with intelligence in their study of subatomic particles. Physicist Gregory Matloff at New York City College did a paper which has gotten a lot of attention in a concept called panpsychism where it is believed that the mind is everywhere. Omniscience present all throughout the universe much like animism with the concept that the spirit is everywhere. And he's not alone, many others believe in the universal mind Australian author and philosopher David Chalmers is one of the proponents of this concept.

Chapter 9
Determinism vs Chance

If you believe Einstein then you know he didn't believe in free will and if you follow his thinking of time as a dimension then you will see.

Along those same lines of thought there are the religious considerations like freewill vs determinism which also comes up in science. The idea that these synchronicities could happen no other way although may seem strange to us. Now I know many may think that this has nothing to do with synchronicities but some things may mimic synchronicity since we are dealing with time space and the universal intelligences playing tricks. But if time could happen no other way then neither can these events which are part of your timeline. Einstein and many other scientists saw our timelines the way the mathematician Hermann Minkowski did. That is Minkoski space or block time. Where your whole life could be mapped out on a timeline just like all world events that have occurred. You could go from past to future laid out but you couldn't change what has already happened.

This of course is because it's a dimension. If you could travel back into a spacetime dimension which many believe man will be able to in the future, it would not change. You could go back and see yourself as a child in the rain walking through a puddle of water and the raindrops would fall exactly as they did. The same place the exact same ones making the exact ripples. My idea is nothing would change and you would have those same synchronicities and they would seem like they were planned. Like Déjà vu and even Déjà vu seems like it happened before because of another theory I have concerning theoretical physicist, John Archibald Wheeler's theory on retrocausality. It's the idea that the future can have a causal effect on the past.

Many psi or paranormal abilities like ESP may be just mistaken identification. Parapsychologist J. B. Rhine came to this exact same conclusion back in the 1930s and built a zener card machine to see if ESP was being mistaken for clairvoyance. ESP may appear to be just a glitch in the timeline. Like Déjà vu, ESP, some synchronicity and many things that you believe will happen and do may all be because of glitches which has more to do with quantum physics, brain function and Wheelers theory of retrocausality. This goes back to my article "Communication from the Future" and entanglement. Subatomic particles are in communication and if they are, you may be just remembering your current future because if time is a dimension which happens all at the same time. You will live this same timeline forever. But entangled particles from your mind may accidentally retain and or communicate some of those memories from the future from your own entangled life. Everything would seem like it was supposed to happen now like it always has. And like I always say "this is why I love science."

Don't get me wrong because I love the philosophy of religion too.

The Definition for determinism is: *"the doctrine or belief that everything in the universe, including every human act is pre-determined and there is no freewill."* After much thinking I have to believe that I'm a compatibilist, meaning that I partially believe in determinism because of so many things that happened to me that seemed like they were planned. I believe many events throughout world history were predetermined or preordained. So to me, it's a mixture of both determinism and freewill. And even in much of what I say as well as many of the others, some of it seems contradictory to what they say they believe. That's the nature of this whole topic people may change ideologically as they get older. Or as far as St Augustine it may be because he had an obligation as he got older to say we have choices. Otherwise we'd have to call everything God's will good or bad. Taking away freewill!

Now I'm not going to say it was God but I don't know if it's from science or some creator entity. If you believe St Augustine or many others who were theological determinists the mechanisms are the same.

So pushing science aside for now, some events have to play out in certain ways and the main players or actors are steered for those particular events. This means that nothing can be allowed to alter those events or timelines of the actors. However for nonessential outcomes, it is my belief that freewill is allowed.

Theological determinism says that God determines everything that happens in the history of the world. And although there's an argument for freewill the argument has and will go on forever. With much debate among theological determinists like Thomsas Aquinas, St Augustine, John Calvin, Leibniz and others. The argument is if whether of not this is a strict determinism and if or not there is any room for free will. We already know by the people we talk to that people have strong beliefs on both free will and determinism. I always bring up these same scientific or theological arguments and always get resistance. Of course I could argue either side but I always play the counter to whichever side the argument is favoring.

Chapter 10
Some Interesting Synchronicities

Some of the damage on the beach next to the field below where I saw the UFO years later as I was writing about it. Synchronicity?

Now let's move on to an example of chance, determinism or synchronicity again with a little paraphrasing: Dr. Tom Leonard, professor of Statistics, University of Warwick, Coventry, England tells the story of a new professor of Statistics giving his first lecture. He flipped a coin to demonstrate the 50-percent probability of it landing either heads or tails. The coin landed on a polished floor, spun around a few times, and came to rest vertically on its edge, whereupon, the class came to their feet in thunderous applause! The odds of this occurrence have been calculated at a billion to one.

This incident might seem to have no significant meaning, therefore not qualifying it in the strict definition as synchronicity. However, we have no way of knowing how the event affected the lives of the instructor or the students. It's possible that the improbability of it may have encouraged a student to pursue a particular field of study or to approach all theories with an attitude of discretion. Thus, it would only have been meaningful to that individual. Synchronistic events may occur that way, by involving several people, but only representing a meaningful relationship to one.

Remember, it only has to have a meaningful relationship to one person. That person would be the subject of the coincidence or synchronicity. But because it was a demonstration to a class by a professor of statistics about the outcome of a coin toss I find it profound and would logically increase the odds. I personally think that the story carried a greater meaning and the fact that it now reaches the world for everyone, means that others must think so too.

Now there are also many stories of Jung but I think this is the first one I've ever read and found it to be fascinating. It's about Jung's most famous case of synchronicity in psychotherapy. He was with the woman patient who recited a dream she had in which she was given a costly piece of jewelry - a golden scarab (beetle). While she was relating the dream Jung heard something tapping at the window from outside. Jung opened the window and in flew a scarbaeid beetle which he caught in his hand, its gold-green color resembling that of the golden scarab in the woman's dream. He handed the beetle to his patient and said, "Here's your scarab."

The woman, who was highly educated and intelligent, had been resisting dealing with her feelings and emotions. She was very adept at rationalization and intellectualizing. After the scary scarab experience she was able to get to the root of her emotional problems and to make real progress in her growth toward wholeness.

The universe had somehow cooperated in her therapy by giving her a meaningful coincidence. The scarab that tapped on Jung's window was no ordinary bug. It was somewhat rare in those parts. It has, as one writer notes, "perennially symbolized transformation and metamorphosis, the very things that this woman's unconscious was calling out for. It was as if the struggle in her soul had been projected like a powerful movie image into the outer world"

Speaking of metamorphosis and transformation! This to me reminds me of one of my own true stories. I wrote it for another book and I called the experience "The Three Butterflies." I wrote the lines below while wailing uncontrollably in the shower when I heard about a young woman an 18 year old college student I knew named Nermine getting killed in a car accident on the news.

"Yesterday I felt a flower as it brushed across my face, as you walked into the room. And today I felt the sorrow as you walked into the light, as you walked into the light. Nermine, what does it all mean? Nermine!

The very next day I was getting ready to have my picture taken for some promotional shots. I walked out my backdoor that morning just to see the day and while still very much upset. Suddenly, a monarch butterfly started to dive bomb me as I was standing near my back door.

This had never happened to me before in my life. The monarch was behaving more like a bee trying to sting me. I was swatting it away but it just kept on coming at me, so I went back in the house thinking how strange that was.

A couple hours later when it was time for my photo shoot I walked out to my car and there was a monarch on my driver's side door handle just sitting there moving its wings. I actually got in from the passenger side so I didn't have to deal with another angry monarch. Now I started to wonder if this had anything to do with my crying and the song I was singing. Maybe it was a message from Nermine.

I then went on to go meet up with the photographer friend of mine Jesse that would be doing the shoot. Coincidentally he was Nermine's friend and that's actually how I knew of her. So, he asks me where I want to take the photos and I chose the shore near where I was then living. I parked the car down at the beach looked around and said let's go shoot under that tree. When I stood to pose, Jesse looked down at my feet and said "There's a dead butterfly at your feet." I hadn't told him about the butterflies I had just seen behind my house at the end of my driveway so it was weird. Anyway, I reached down to where it was under some blades of grass and picked it up. I realized it was still alive and I said I want you to take a picture of it leaving my hands. I opened my hands and it flew away as he took the photo. I then told him about what had just happened with the other two butterflies.

I now to this day think that this was Nermine giving me a sign, knowing how much her death affected me and in some way saying "I'm OK don't cry." Also to this day every time I think of her wasted young life, my eyes well up. I could have chosen any place to take those pictures and once I got there, could have chosen any spot. Once I chose the tree, I could have chosen anyplace beneath it to stand. So what on the surface appears as a bunch of random choices turns out to have comforted me. Is there life after death? I don't know if it will be in anything but a spiritual sense, but reality is stranger than fiction. Or as main stream science would say, three monarch butterfly coincidences.

The story about Nermine is true, but the events with the monarchs happened the next day after I heard the news. So what was her soul, spirit or mind still doing here? And if it wasn't her, then who or what was trying to ease my anguish? It's not like there were swarms of monarchs flying around that day, I only saw three.

Synchronicity and the Secret of the One

Then I heard some psychic lady on TV talking about butterflies as meaningful messengers from people who've crossed over which got my attention and it was Sylvia Browne the author and so I read one of her books. As it turns out from more than one source, butterflies are very spiritual and carry a lot of significance as a sign of working with angels. Many people report seeing butterflies shortly after a loved one's death. As they are supposed messengers of the departed.

In my life it seems odd enough that I continue to have these synchronicities giving me something to write about as far as the odd or paranormal events go, but the synchronicities are many. I chose a few as examples because like the first story I told you or the light most of the rest people would not believe or are too mundane to include which would lead into hundreds more mundane stories.

There are many stories of chain or compound synchronicities abounding throughout the world by countless people. There are books and many more never written down many more I'm sure not even recognized. Everyone has some deeper meaning to the one recalling it. I myself have my own with some so dramatic I chose not to write them. Some might even believe I was stretching for meaningful interpretation but I can assure you that I wasn't digging for meaning at all. In many I had to stop myself because some had too many meanings. What came out from the universe were the obvious things that gave them the meaning I recognized. Just like the colored lights, the storm and the washed away and destroyed houses.

Another I wrote which was edited out of the article by a website editor. I think I may have decided to leave it edited out permanently from the article. Which may lead us of course to the power of the mind to actually subconsciously feed into the outcome. Yes, or more accurately "intention" can also be a very strong force in everything we do or "will." Like the "ONE" or as the other half of the equation we sometimes have to guard our thoughts. I often purposely try to prevent negative thoughts from somehow involving themselves. Or like the original scientists who realized it was affecting their experiments. Or Wolfgang Pauli or those around him who realized how intention (good or bad) can affect the outcome.

"Always think positive" it's a good motto to live by!

Chapter 11
Conclusion What About the Numbers?

Well we did come through this and all these revelations later I thought you would have gathered the meaning which should no longer be a secret by now. Is it scientists or sorcerers, shaman or gods? It should have revealed itself to you by this point. Many things are just as plain as day. There's a sun in the sky which gives our world meaning when we see the day. So much of everything have simple meanings for us to see. No sun, no light, no meaning – no universe filled with sentient beings, no meaning. I'm not trying to follow the leaders with scientists like Einstein, Bohm, Plank, with their beliefs in a super intelligence in the universe or trying to emulate leading philosophers throughout history - Kant, Pythagoras, Aristotle, Plato... The theologians like Aquinas or Augustine they knew that the numbers are the meaning. Just like with Jung and Freud or Jung and Pauli it's the interpretation by the subjects that's more revealing. When you walk by and notice the numbers with corresponding perfect alignments. Then I do and we both have different interpretations of perfect that also tells the creator a lot. Unless it is the God entity because he would already know what the meaning is.

Everything is the meaning as a whole and it is all meant for us personally but to the universal intelligence it already comes with a meaning. But I always had my own belief that there was a superior source intelligence and really I don't care how anyone interprets it, but I know it's there. Now we can share what it meant to us personally like I did with my number which is three, but it is no longer a secret. The one is the one no matter what you choose to call it. All these synchronicities have meaning. Just like all these numbers do every time we see them we can get to decide for ourselves. That's what the intelligence is telling us. That's the other half of the secret – that we get to decide and we aren't told, we have to figure that out for ourselves. Just like we have to decide the meanings ourselves because half of it came from our own minds. I mean we wouldn't need a co-creator of our own universe if we only needed the ONE Source Creator!

Synchronicity and the Secret of the One

Synchronicity and the Secret of the One

Resources

The King James Bible.

Albert Einstein (2014)

Charles Fort author (1917) "The Book of the Damned" (1925) "New Lands" (1932) "Wild Talents" Dover Publications.Mariner Books.

Carl Gustav Jung (2010) "Synchronicity: An Acausal Connecting Principle" Princeton University Press.

Sylvia Browne (1974-2008)

Wolfgang Pauli

David Bohm, (1980) "Wholeness and the Explicate Order" Routledge

Erwin Schrodinger (1926) Schrodinger Equation,.com (2014)

William A. Tiller, Ph.D. and **Walter E. Dibble, Jr., Ph.D**. (2009) "White Paper on Intention"

Fr. John A. Hardon, S.J. (2000) "Thesis X. Original Sin" www.TheresAPresence.org

Dr. Tom Leonard, Meaningofsychronicity.com (2014).

Michael Talbot (2011) "The Holographic Universe" Harper Perrenial.

Rene Descartes (1641) "Meditations on First Philosophy" (2011) Simon and Brown.

Rupert Sheldrake (1988) "Morphic Resonance" Times Books.

Sin Thesis Real Time (2014)

Saint Augustine of Hippo (397 - 426) "On Christian Doctrine," "On Free Choice of the Will"

Saint Thomas Aquinas (1400 c)
Thanks to many others for so many wonderful articles

Robert Torres author "Synchronicity and the Secret of the One" © 2018. All copyrighted material used in this book falls under the "Fair Use Doctrine" as provided for in section 107 of the US Copyright Law. In accordance with Title 17 U.S.C. Section 107. Much effort has been made to cite what I could within the bounds of space. Much more material was researched for this book than was referenced. I apologize to anyone that was left out of these citations. If you have any questions, please contact the author and I will attempt to verify and add a citation on any further resources used for this book. **Robert Torres** Email: Rttowers3@gmail.com. / www.facebook.com/robert.torres.94064176Torres or Amazon for purchasing info or questions.

About the Author:

Robert Torres aka Bobby T is the author of several books including "Synchronicity and the Secret of the ONE" and others using a pen name. He is a published singer-songwriter of "Then There Was Rock" and artist. Torres has done work for many well-known national US companies as a freelance. He has also worked as a magazine editor, technical, music and content writer, consultant, web designer, promotions, computer multimedia technical production and the list goes on. Torres writes as a freelance contributor for many online sites and offline magazines, like: New Dawn, The Epoch Times, Mindscape, Phenomena magazines. Truth Theory, The Mind Unleashed, Wake Up World, Enlightened Consciousness, Learning Mind, Disinformation and many others.

"I exist on many plains of reality simultaneously, one of many, is one of MAN." — Robert Torres

This book "Synchronicity and the Secret of the ONE" was written using excerpts and concepts from the non-fiction conceptual article "Synchronicity and the Secret of the Co-creator" and is now available at Amazon written by the author Robert Torres.

www.ingramcontent.com/pod-product-compliance
Lightning Source LLC
Chambersburg PA
CBHW070203230526
45471CB00002B/798